To our readers, those students just beginning a life of adventure and learning, always remember:

You are braver than you believe, stronger than you seem, and smarter than you think.
~
A.A. Milne

Written by

Brent A. Ford

&

Lucy McCullough Hazlehurst

© 2017 by nVizn Ideas LLC

www.nviznideas.com

Well, will you look at that?

Well, will you look at that?

Well, will you look at that?

Well, will you look at that?

Well, will you look at that?

Well, will you look at that?

Well, will you look at that?

Well, will you look at that?

Yes, what a prickly tail. Watch out!

But why?

Well, will you look at that?

Yes, a long and fluffy tail.

I wonder why.

Well, will you look at that?

Yes, a tail that can get quite fat.

I wonder why.

Well, will you look at that?

Well, will you look at that?

Yes, what a lopsided tail you have.

I wonder why.

Well, will you look at that?

Well, will you look at that?

Well, will you look at that?

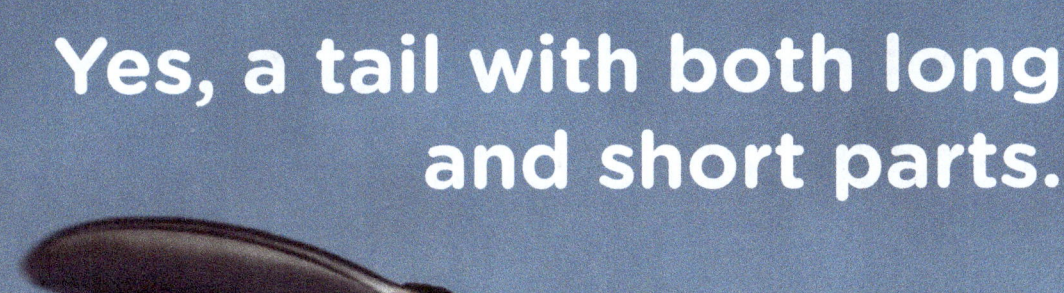

Yes, a tail with both long and short parts.

I wonder why.

Well, will you look at that?

Well, will you look at that?

Ozzie & Alina Adventures

Updated Classics

Adventures from nVizn Ideas

Science & Nature eBooks

Interactive eBook

Variety in the Animal World

Well, Will You Look at That?

Well, Will You Look at That?

Well, Will You Look at That?

Well, Will You Look at That?

Well, Will You Look at That?

www.ingramcontent.com/pod-product-compliance
Lightning Source LLC
Chambersburg PA
CBHW081205020426
42333CB00020B/2626